From #1 National Bestselling Author
C.V.Conner, Ph.D.

Mom
Help Me
Tell

MY
STORY

A Daughter's
Courage to
Let Go

Story by Alice August

Mom Help Me Tell My Story

The Holly August Story
a walk to remember.

Alice August & C.V.Conner

COPYRIGHT NOTICE

A Message From The Co-Author:

Holly,

You've said before that I'm more like a good friend than your uncle. That touched me deeply.

I've wished I could have been there for you even more. But since heaven makes that difficult, I thought that this would be a way of sharing with you all the things I would if you were still here – in the spirit of a good friend.

As you grew up, went through life experiences, you learned things, a lot of them the hard way. You went through struggles, you got knocked down, you got up and you learned to deal with life challenges. Life taught you all along the way to heaven. And these are some of the things I also learned in my journey so far – what I've found to be important to happiness. So much of what I've learned though, I learned from you.

"A smart person learns from even the most unexpected people and events."

This book is filled with memories and thoughts of you. You are beautiful inside and out. And you were never afraid to be who you were and let your inner light shine through.

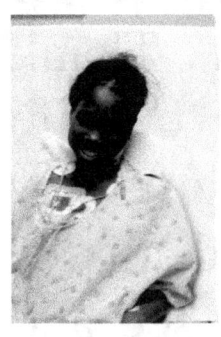

We all loved that, we all miss that, because you *ROCK!*

Love
Uncle Beddy.

Message From Alice C. August

A Promise Made my Love is a Promise Kept
Mom will you help me tell my story?

This book is dedicated to the life and memories of my beloved daughter Holly Alyce August. Thank you for staying as long as you did, sharing your life and your desires and especially your love for children and God with me. Love you and misses you, but knowing it was worth the trip to see and know heaven in reality of what we talked about keeps me going. See you when I get there.

INTRODUCTION:

This book is to share Holly Alyce August with the mashes per her request. Born with improper development of the small intestines she had several operations that resulted in IBS. Not knowing about IBS caused many years of not having proper treatment. And by the time I come into the knowledge and doctors discovered what had happened years of her life had been shorten. She asked me some months before her passing to tell her story and I am and will until I have no one else to tell it to.

If you are reading this book please pass on this information and consolation to anyone that might or know someone with this problem. Early detection and knowledge is the key to avoiding much hardship and shorten life.

Chapter I

It all started in November 1986 while enjoying life and anticipating going to ATDS Trucking School in Waco, TX. Having dreamed about being a trucker for a large part of my life, finally I was off to Trucking School. During this time I was living in Dallas, TX with a friend. She made the statement a week before I left, "What would you say if you knew you were off to meet your husband?" I just laughed and kidded with her the way we always did about me marrying again. Needless to say two weeks after failing trucking, I was dating the man I had met in trucking school and he had asked me to marry him during the first week of being there. A month later we are engaged and married. It all happens so fast I forgot what my friend had said. He lived in Houston and I returned to Houston after failing trucking to continue education at the University of Houston (Computer Science Major). Once back in Houston I contacted him and we were inseparable from this point on. We were married January 13, 1987, and Ms Holly was born July 14, 1988.

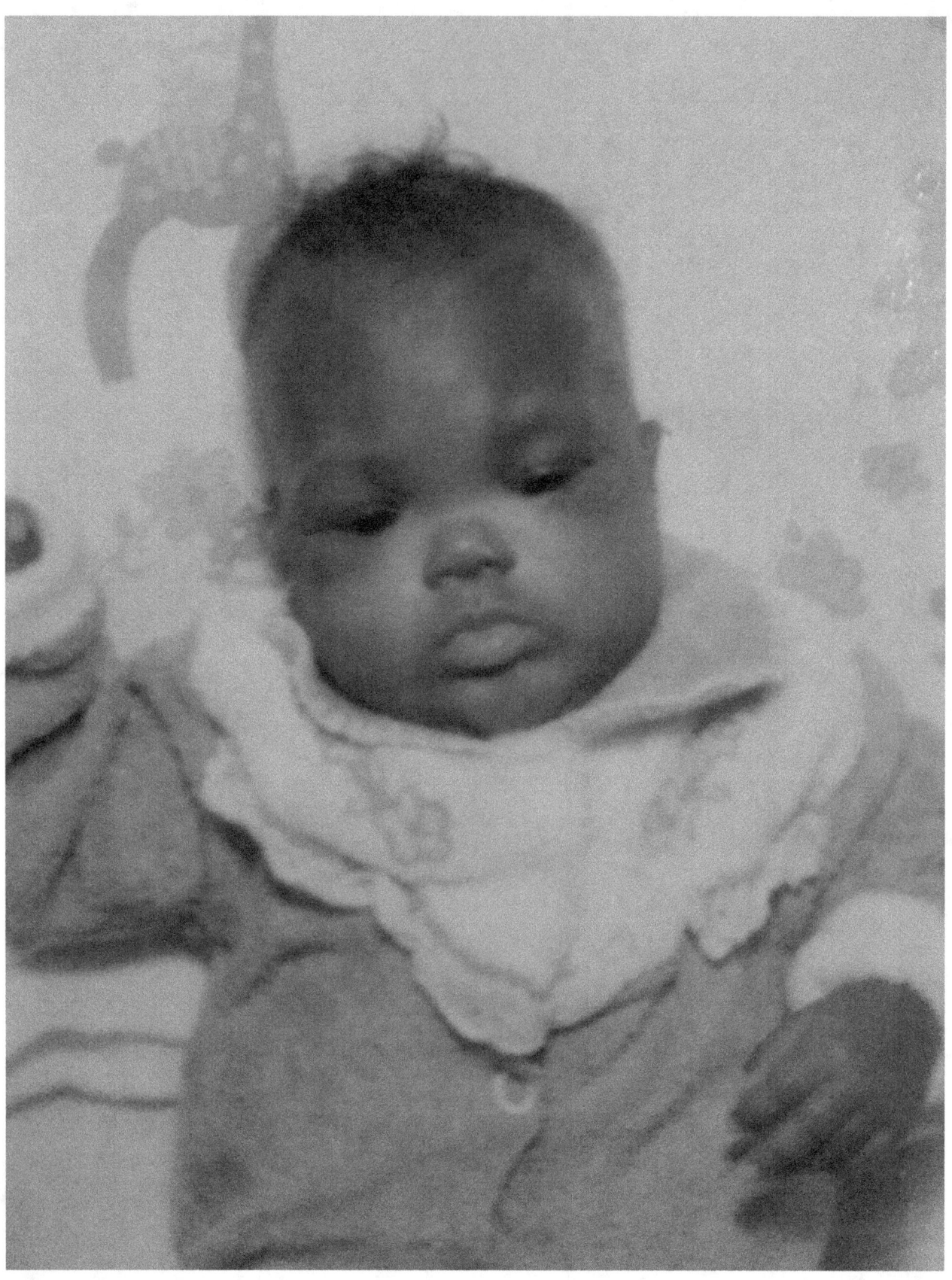

Howard and I are playing and wrestling when I feel like I have to use the bathroom. He's holding me thinking that I am kidding, when the expression on my face lets him know otherwise. As I enter the bathroom water pore down my legs as I try to stop the flow to get to the commode. "Howard, I am peeing on myself", he yells from the room, you had better not be peeing on the floor with your ole but". I re-assure him that water is coming down my legs and I can't control it. He comes to the door and realizes that my water has broke. This is my first child and I know nothing about water breaking other than old women talking about it terms to avoid children understanding the true meaning. He turns and runs out the front door upstairs and gets our neighbor to help get me ready for the ride to the hospital. He's nervous and excited at the same time. Me I am wondering why I am not hurting or going through all that I have seen on TV. Needless to say nothing has been the story of pregnancy I've heard down through the years. I haven't had a good seven and a half months. Everyday from conception to now has been a wrestle to keep her safe until delivery.

I discovered my pregnancy while having a severe toothache and going to the dentist for relief. As she's checking me out, she refuses to help me and sends me to my primary doctor. Again ignorant to what's happening, I go and the doctor checks me out and tells me I am pregnant. This comes as a shock and I just look like a deer in headlights, but when my husband hears he's ecstatic and full of ego boosting joy. Nobody will help me with my toothache, so home remedies and finally a dentist that gave me antibiotics to get the infection down to endure until things could be corrected.

Now the GBYN visits started and during checkups we discovered that I have fibro tumors and will have to be extremely careful in order to keep the baby safe. Food was almost impossible to keep down; Oysters and cherry peppers were the only things that I could tolerate without vomiting. I'm willing to endure whatever I have to keep my little princess safe and sound until delivery. From one visit to the next up to this time is very painful and I am on bed rest. Howard's the perfect husband washing, cooking, and carrying me from one point unto the next in the house and the doctor visits. He sits through every one of them taking in everything said. He's determining to be a father to this little person. He talks about Howard August III, but I know she's a girl and I have the name and colors already in mind. We argue back and forth about names and gender.

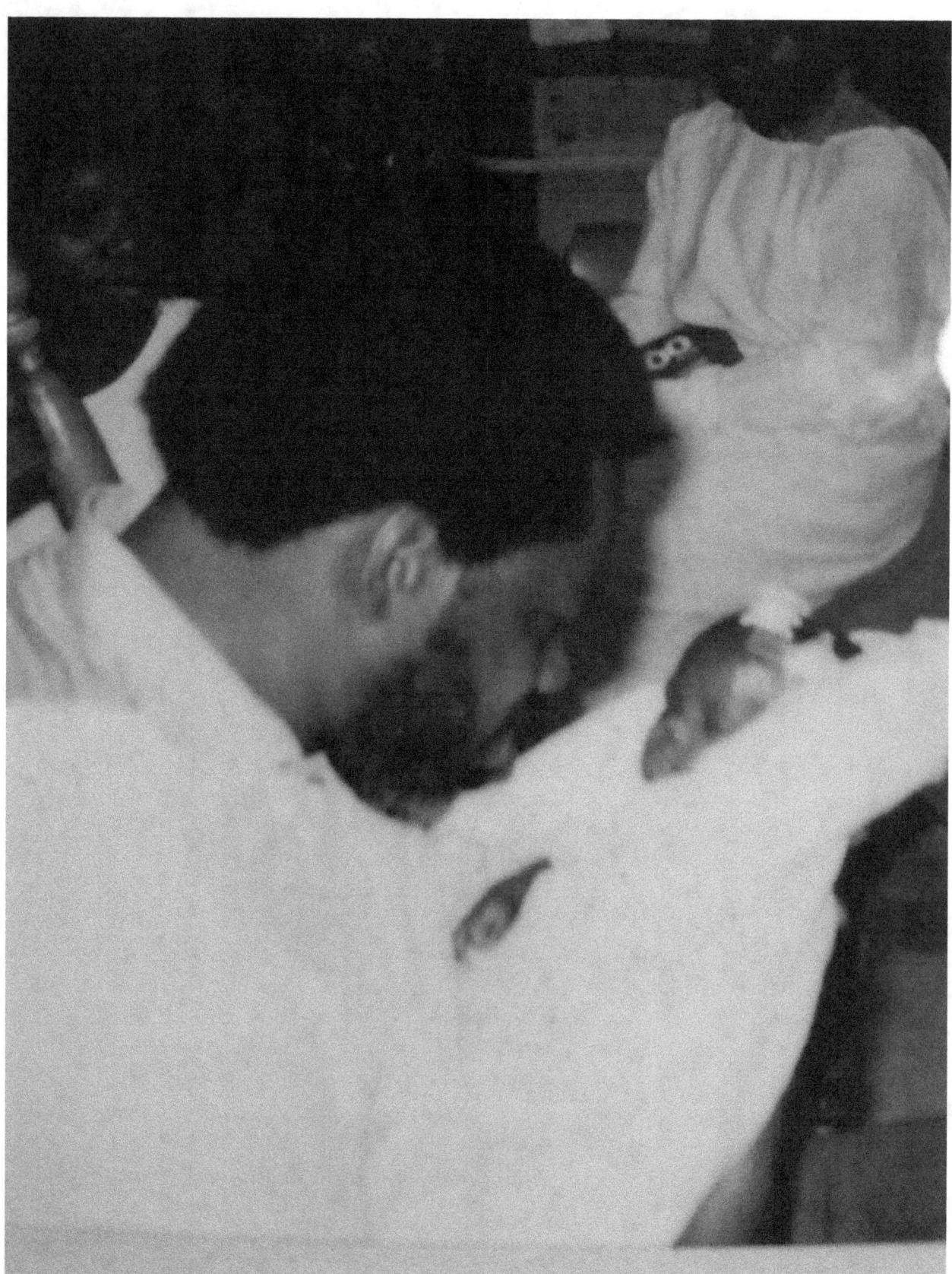

Chapter II

This is where her journey began. Born prematurely she encountered problems with her intestines growing like sausages. This caused her to have three surgeries to correct the problems before she was 6 months old. Each time doctors didn't give her any hope for surviving. She lived in ICU at Memorial Southwest Hospital /Houston, TX for the first four months of her life. We never missed one day of visiting every visiting hours.

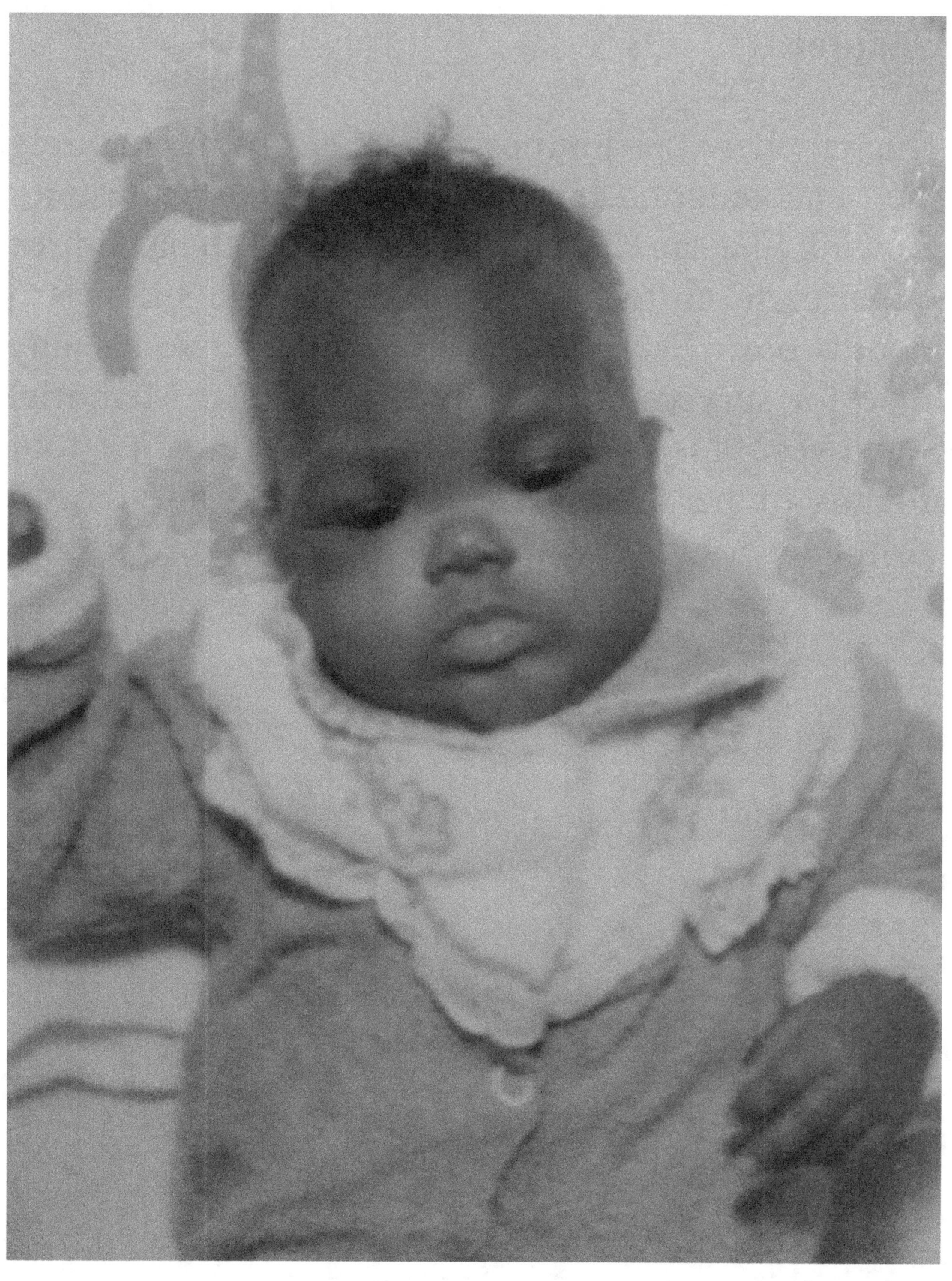

We were allowed to hold her and the immediate family joined most times. Having intestinal problems didn't allow her to take feeding; therefore, she had a feeding tube. The most beautiful little face and eyes one can ever encounter. We sang songs (Yes Jesus Loves Me) etc. During this time I watched her dad hope against hope that she would make it home ok. It is his second chance on being a father having not been with his son during his years growing. He wasn't even sure where he was, though he longed to know and see him someday. Very few conversations on him just knew his name is Jeffrey.

After the four months over and her intestines put back inside successful and she has a bowel movement we all celebrated like it was us. Dr. Robert Bloss, surgeon had a field day celebrating. Needless to say we did too. He is and was a God sent into our lives along with Dr. Joan Sterner, etc... Now she's reached the milestone needed to transfer to Texas Children Hospital for further aid in getting foods that her little body will tolerate. No feeding tube, sucking her bottle independently. And you know it her dad ran into the hospital to be the first to feed her. I am so glad for all that happening, that just knowing that she's eating is enough for me.

Days past and arrangements in place; the ambulance arrives and off to another hospital to see our little angel excel. We are here and things go well it seems until a nurse gives he the wrong dosage of medicine. The monitor stops the priest comes in and we are dead in our tracks. When in comes a doctor that asks for the medical record and he looks goes to the medicine cabinet in the room and takes a serine and fill with a solution and inject it into Holly's stomach. The lines start again, and we start breathing again. When the doctor exit the room the nurses ask who he was and nobody has ever seen him or have any idea who or where he came from. I know God prepared an angel for us in the midst our storms again.

Having this encounter she is released from the hospital within 2 to 3 days following. She goes home in time for her very first Christmas. Her grandmother is waiting in Beaumont, TX with decoration and all. Her great-grandmother is there too. Granny (great-grandmother) a Kidney patient had made the trip to Houston, because it was taking too long for her to come home. She was well up in age and failing health, she wanted to see Howard's child just in case Holly or herself didn't live. We had a celebration to remember until Grand-mother and Auntie accidentally pulled the center line out trying to change her diaper. Seeing the leak around the line, we had to leave early and head back to Texas Children Hospital.

Once there one of the surgeon gave her the nickname "Cat" because everything they said she couldn't do she lived through it. Dr. Quavers, at Texas Children Hospital, Houston, TX decided to allow her to try it on her own without the TPN pump December 1988, when it malfunction and started leaking. She did great and made her regular visits with the regular changes expected for young babies. At this point we begun to focus on the other physical ailments caused by the tumors affecting her right leg in the womb she had corrective surgery to correct a club foot from not having enough room with the fibro to hold her foot in correct position.

During this time my marriage fell apart under the pressure of caring for Holly. This took a toll on both of us. I was into working and caring for us all and his jobs weren't holding up long enough to have two insurances, so I had to work hours that accommodated care taken and work. He spending so much time along to think and drift backs into substance abuse tore us into pieces. Love never died, we just had to many things against us and not enough time to mend the broken pieces before other ends ripped apart.

Divorced and still trying to keep Holly safe was our focus now, and if I must say so we did a good job at it. She grew up and understood more and more as she did that our choices was for all of us. We worked, lived and recreated together as much as possible. Sometimes we even tried to get it back together but again sickness prevailed. After being released from most of her doctor's care and growing now, walking and talking it's time to move to the next phrase getting eyes, legs corrected. Lazy eye, Clubfoot. The clubfoot is taken care of by Dr. Douglas Houston. A success but by now doctor bills are mounting and insurance companies are investigating spending.

We are directed to the Scottish Rite Hospital in Dallas, Texas. At this point her medical bills are 200 dollars short of one million dollars. We spent two years in braces and splints (before she is able to walk and run on her own). This was her story for many years to come. She started out with being able to go home without a bag on her at the age of six months old to wearing only a TPN catheter and a TPN pump. Her victories over one hurdler after another are very trying but a joy for me to see her through. Not knowing what it is to see her without a testing attached we go into the habit of focusing on her favorite scripture and looking for all to be glorious each time "I can do all things through Christ that strengthen me".

Holly was short and small in status when you met her, but once she opened her mouth her status and size no longer matched. She had a heart for people and love that shined at a rate that there was no denying. Not to mention the smile that she wore each time she greeted someone. She was looked upon as retarded and slow in appearance also until she opened her mouth and heart to you. Needless to say it was very real to her, and she would often ask me. "Mom, what's wrong with me? I always assured her that it wasn't her but the ignorance of the ones viewing what they had no idea about. This didn't make it easier for her I learned as time progressed. She tells me stories of setting looking out of the window at the babysitter and having the sitter tell her one day you'll be able to go out and play too. Also people would question her legs one smaller than the other. It breaks my heart into little pieces but I can't show it. I have to endure for her sanity. It begins to help when her uncle Joel shows her his leg in brace and tells her war story of how he got his leg hurt. They both walk with the same limp and similar brace. They are ever so close and adoring to each other. He teaches her how to walk and adapt to the brace, this cause a change in her that lasted a lifetime. Now no one can tell her anything about her legs for they carry her just like everyone else. Serving the same purpose.

Off to school and an aggressive attitude that she's developed from the past hurts and disappointments, She's a force to be reckon with. Living the in your face and front and center personality of hers. Often coming across as arrogant until she finishes speaking. She showed love, respect and empathy on a level that even I had to re-visit. By this time I am realizing what life has done to her for both the good and the bad.

I make it my business to take her to and from school, to avoid her being injured by falls or mean little children trying to earn a reputation. She loves it we have the morning breakfast and ride to discuss our day. She is out and out of sight catering to leadership, picking the teachers that have a heart and the ones that make a different based on appearances. Now having been around persons with great insight and spiritual connections I am seeing a side of my daughter that gives me to know that she's is God sent for this time in all of our lives. She will touch so many lives before she leaves.
First time I notice this is when she would volunteer and pray for people in church. She always loved to

sing along with the choir and immediately after church she made her rounds and hugged different ones. On the way home she would tell me to pray for certain ones because of different things she felt and was told in her spirit about them. In time different ones made statements like "that Ms Holly is something else, once you get to know her". She traveled with our church group sometime, and visited with family and they too begin to notice her God given abilities. So much so that varies ones would ask her for her opinions on situations. The answers would always be on key.

Kindergarten was a major testing within itself. The teacher tells me that she's special and needs special education and needs to be held back a year. I refused based on what I know; she needs a teacher that understands her as an individual not an appearance. So I go into action (after spiritual counseling with our pastor) I refuse to hold her back and she goes on to the first grade. Entering first grade she's excelling and keeping up then her school records comes to the school in Houston from Mississippi where she started and the suggestions of special education. One meeting after another and I refused she goes through first, second grade in public school with the same thing from year to year. Now I have had it with them and their foreseeing of future problems. I pulled her from the public school and placed her in The Varnett School in Houston Texas, this is a motivation to her self-worth. This school is outstanding and the curriculum is awesome. She loves the school, teachers and the children looks forward to school daily. Starting third grade is pleasant for her and several teaches and office personnel have become her favorite people. She holds conversations and shares on the way home. The field trips are outstanding for the parents, teachers and the children's. I see progress on a level that keeps me encouraged for her in the future. Then it starts a new teacher come to Varnett and on the way home Ms Holly says to me "Mom, why my teach can help everybody but me?' I asked her to explain what she means. Well when the other kids raise their hands she takes their questions and answers, but each time I raise my hands she tells me to take it down! Shocked, I assure her that I'll take care of this in the morning. She's relaxed and trusting knowing that's what Mom does look out for her well being. We finish up the day with

homework and dinner. Her favorite pass time is and
bedtime stories.

We arrived early and made ready for the day and off to school. Upon arrival I leaves her in the lobby with the assistant principal and goes in to talk with the teacher. She greets me and we begin to talk, note she's not contacted me or sent any form of information regarding this. I speak and introduce myself and tell her what happen in the car on the way home. She looks at me and sigh, " I don't want to loose my job". I assure her I am not here for your job, but my responsibility is to my daughter. She asked that we talk off the record. I agree and she assures me that " Don't want to say she can't learn, but I am not the one to teach her". I thanked her for her honesty and proceed to office and make arrangements to have Holly pulled from her class and placed into another teacher's class. I am told that this is against school policies. I leave Holly at school today with the assurance that I am working things out and today is the last time she'll have to deal with this issue. I asked to talk with the principle and was told I wouldn't be able to do so. Leaving in confusion as to what just happened, I begun to pray, "Dear God, tell me what to do." I get home and pulled out the yellow pages and begin to look for a school for my daughter. Public school seems to have special education of the brain, Varnett School though a great school I can't get the information to the right source. As I am looking at the school listing this one particular school just stands out on the page. I know this is my prayer being answered. I am off to the The Cliffwood School (School for the learning disable). I go and meet the Principal and owner, give her my story. She listens and understands, sharing her story of why she open the school. She has a son with a disability that schools used to judge and place him. This prompted her to open this school for him and others like him. I was told to bring Holly to the school on Saturday to meet with her and tour the school. It's Friday and I leave in awe can't wait till tomorrow. Thank you God you've given us the chance of a lifetime. Its three o'clock picked Holly up and shared that news she's glowing. We both could hardly wait till tomorrow to visit the school.

Saturday morning and we are off to meet with Mrs. Weinberg, owner of The Cliffwood School. Holly meets with her and tours the school. She comes out of the office after the visit and runs into my arms and "Thank you Mom". Cliffwood goes from Kindergarten to High School and College assistant if wanted. The following Monday I pulled Ms Holly out of the Varnett School. This prompts the assistant principal to notify the Principle. I talk with her and assure her that the best thing for Ms Holly is The Cliffwood School.

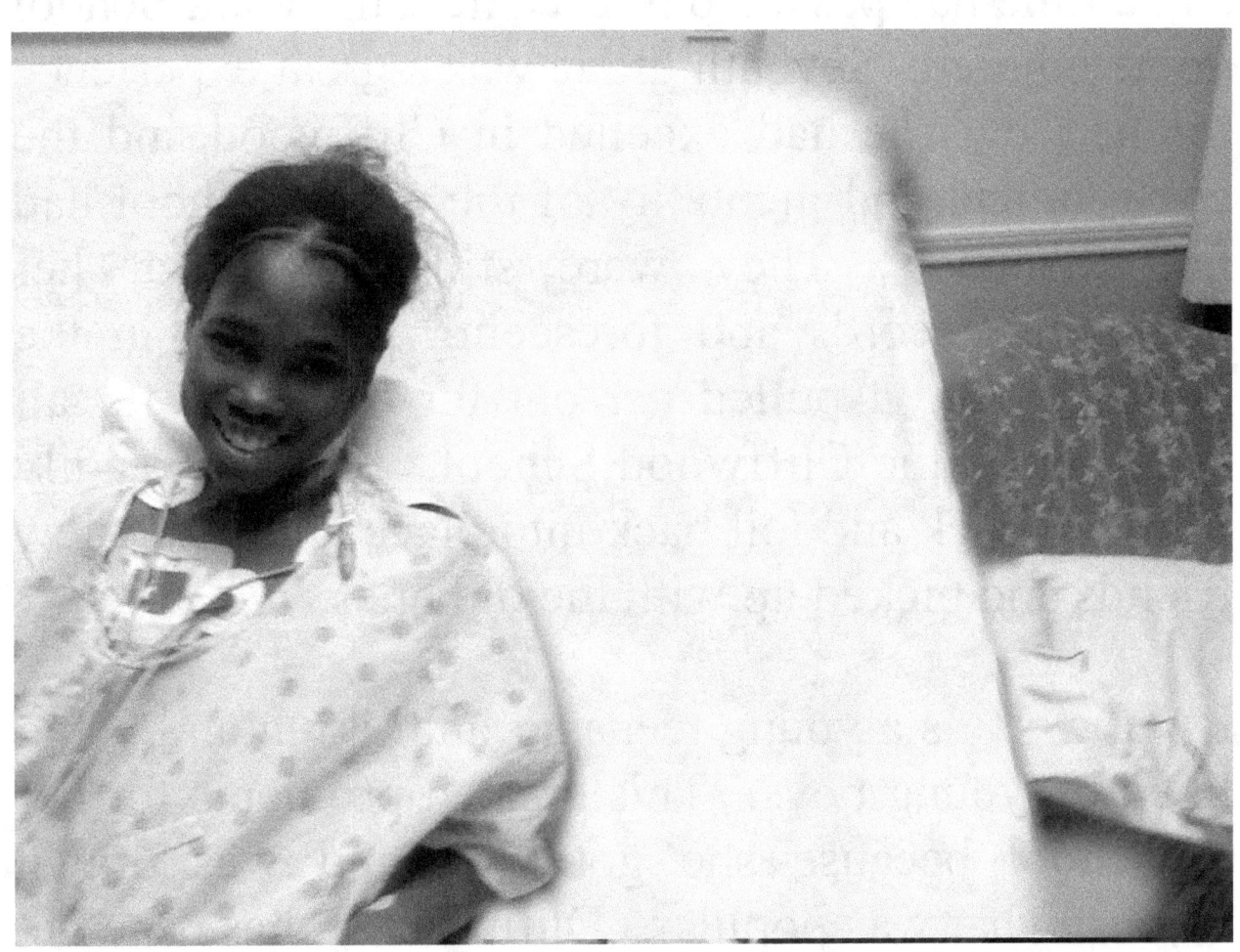

The principle apologizes for the way things happened and tried to make it better, however, I am a firm believer that all things work together for our good. Holly started Cliff wood School and excelled until she was in the seventh grade. Her Dad wanted her to get back in the public school in order to learn how to deal with what he called the real world dealing with life the way it really is. I gave it a try since time has passed but left The Cliffwood School doors adjured. Seventh grade was a great experience for her, but she had excelled in Cliffwood and the environment and mentality of the public school had not changed. They were still looking at her beginning records and foreseeing problems in the eighth grade. I pulled her out for once and for all and back to the Cliffwood School. She was readily received back and fell back into place and made new friends and picked up with the old ones.

By now she's a young teenager and liking and being liked by young boys. This created problems for her in school because she got with and was drawn toward what was popular. During this phase she get in school suspension of course I am called to parent teacher meetings. Now this isn't the first time; I have had it its time to call in the cavalry. I called her dad and shared what's happening. He and I are separated at this time in her life and this contributed to the problem.. We remained friends and did things together as a family. Both of us had come from a split family and knew the consequence of the after effects on children. But we were determining to protect her from the full blow. She's writing letters that inappropriate and on the phone and texting inappropriate things; breaking the rules taking the phone into the classroom. I do not understand this, going from an ideal young lady to a person who bends over to be accepted and admired by the wrong group in the school. By this time she's in the eleventh grade only one year to go. We are on a good path again and she's excelling and started dating her little friends.

It's an awesome time in her life seeing her blossom, live and love it. I discovered her dad is sick with heart disease, she knows this and it's taking a toll on her in a devastating way. He talks with her and gets her to understand that she'll need the education when he's going to be fine. She listens and promised him she'll finish school.

A year comes and past she spends as much time with him as possible. She cooks and talks with him often, holding on for dear life to the daddy that's been there for her in every way possible. A love that is so great that it's visible to all that encounters the two together. She loves her mom but she's daddy's girl. If I could lessen the pain and fears she's feeling I would but this is our cross to bear separate and together. At this point her dad gets worst in health and we are called to the hospital. He has triple bypass surgery and dealing with other illness he goes through it successfully, but not without other complications. He's admitted into a nursing home for rehab. This sends Ms Holly into depression. I watch and help as best I can now she's taking on physical illness due to the depression. Oh God, help where to from this point. It's a Friday I pick her up from school and the teacher tells me to have her checked out she looses her ability to lift her arms and legs at time.

She has experienced this for a long time due to the short gut syndrome as results of the multiple surgeries after birth. This problem with numbness landed her in and out of hospital most of her life, but never to this extends. With this she's back into the hospital and told she has kidney disease. Things gets better she is released and put on meds to keep kidney healthy as long as possible. I get her Houston's number one Neoph0liost team and practice the daily routines. It's nearing her graduation and her dad being in Rehab makes it less exciting because she wants him at her graduation. He's released to go home on weekends. This makes it better to some degree, but not the same they look at each other with so much desperation. I stay strong for the both of them and watch and pray for a miracle in both directions. She never tells him the extent of her illness she just want her dad to be well. Praise God, he gets to come to the graduation and it's off with a great start and finish. Our closes friends and family is there and she just looks around and breaks down and cries. She later tells me that it was just graduating not knowing if she would be able to. " Mom, so many times I was told I would never learn or graduate and having my dad live to be here is just overwhelming. I did it, Mom we did it. "

We spend many months to and from the doctor trying to get them both in a better place. She excels and then falls of the wagon. But I must say she's a champion to be compared to no other. Her dad's back at home and we are looking after him, he's doing great and making progress. Holly's health bounces back. Dad and daughter back together again.

Holly comes to me and lets me know she's ready to move to Mississippi. She wants to attend the Junior College there in Meridian. It's great for with me because we had plan to move to Mississippi when she finished school. I have purchase a house and a trailer there for the both of us moving there. My sister gets her accepted and paper work in place.

Now the big chore, telling her Dad she's not staying in Houston for school. She is his life at this point. I know this, but Holly needs to get away from the city and the pressure of sickness and not knowing how long this is going to last. We agree to go and tell him together. We arrive at his place; he's home taking care of himself doing great and getting back into the swing of things. If looks could kill, I would have died when she tells him. Immediately he blames me. He thinks it my idea and feels that I am sending her away from him. We let him think that way and assure him that she'll call often and visit every time she gets the chance. He's hurt but perks up for her because she's excited. That's all that matters for me she's happy and trying life on her own. I've lived for this.

Driving her to school and watching her take matters into her own hands thrilled me. She had made the journey successful. Time for a new chapter in her life; our little girl had beat the odds. School was a challenge for her. The campus wasn't what she was use to. But being a small town college was good, she made friends easily and then things got better.

After being there for a first semester her Dad is diagnosed with cancer.

I received a phone call from my sister in law, telling me to be ready. Howard's back in the hospital and they're on the way to get me. I ask them not to tell Holly, I want to be the one to tell her. We visit him and he looks frail and gray, but I don't ask questions. I explain that we are getting Holly on the phone, but not to tell her he's in the hospital. I also explain why, she would go to pieces, not being able to get to him. He agrees reluctantly without a word. Things spending so fast in my head now that I never asked any questions just visited and watched him as he watched me. Later Holly calls me and asks how's her dad? I explained that we were with him in his room when she talked to him and he's good. I am coming to get you this weekend and bring you to Houston to see him for yourself.

With that she calms down I thought, however, hater I learned from my family that she pitched a fit once off the phone. She was angry and worried that something had happened to her dad and nobody was telling her. I do as I have down through the years take her to him. He's fragile and weak, we try to be there as best we can, but by this time he has a lady friend and I can only do a little. The sister and mom try too.

After one semester Holly returned home to look after her dad. We spent every weekend with him and took him to church with us often. Times taking a toll on them both she clings to him as though he's vanishing. One night I get a call, it's him. "Alice, I am in trouble can you take me to the hospital". We take him and spend several days before he's released and returned home. Now it looks like he on the mend. Later on he's back in the hospital and this time, he's unable to talk and it looks like the friends that's been hanging around him has taken advantage of him and not allowed him to take his medicine. Oh, my God, he's in trouble and this time we are in no place to help him. He's released to rehab, but someone drops the ball. Howard passed away and this left a whole in My Holly's' soul that time never touched much less healed. She was never the same. The smile and vigor she had for life left and she tried to move on but it took a holt on her in such a way that her health went down hill. During this time fighting the depression and kidney disease she was molested and this was the final straw.

Holly had made her dad a promise she would complete college and she did. Her health didn't allow her to physically attend, therefore, she did it online. Once she complete her class and got her papers she said "ok, dad I kept my promise".

I watch her go to dialyze three times a week just going through the motion. I watched as tormenting spirits attacked her and she fought for the safety of her family. It was surreal and like nothing I would hope on my worst enemy. She lost weight and evenly her sight. I came home from work one morning to find her lying on the floor at the end of the hallway in blood splashed clothes and asleep not being able to reach her room for confusion. At this time she's keeping it all inside, for my protection cause it's just us now and I know she loves and want to spare me any more weight. I run to her and call out her name and she opens her blood shot eyes and looks at me with a busted lip and rips my heart apart with "Mom, I am so glad you came". I whole her and apologize for leaving her home that night for work". She had asked me to leave her home with the dog in her room and she'd be all right. This is what we had done for years. She's 23 years of age now and very dependable and responsible. My brother who stayed with us after her dads' death was away on a hunting trip. I called the doctor and he had me bring her to him instead of going to the hospital. I never knew kidney patient lost their vision, or suffered as much as they do. However, walking this road to recovery with my baby has taught me what suffering and living is. She keeps trying and we keep praying for the healing and kidney transplant.

Finally she's on the transplant list and going through the process of classes, testing and doctor visit to receiving her living-sustaining kidney. I see hope in her eyes as she talks about it and smiles and make every stride towards the journey. During this time due to the molestation and food and medicine affects she weakening again. Often I have to take her to the hospital for two weeks stays. Here she's met it seems an angel, because they both comes into the hospital at the same time, same age, same phone, and same laptop.

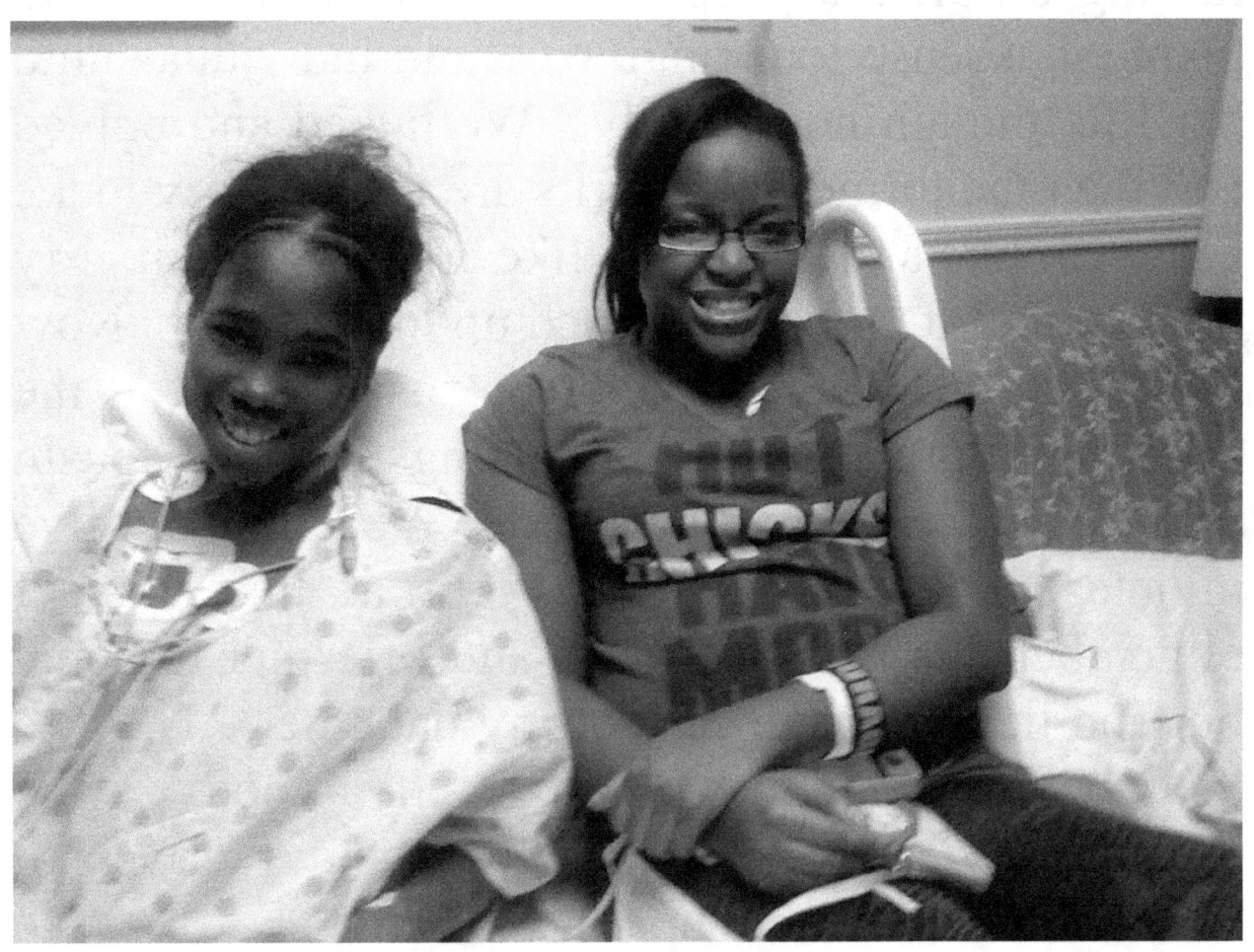

They walk the hospital halls and plays with the young men's and even pork fun at people that'll chase them. It's good to see her smile and enjoy as much as possible. Having begun to have violent spells and not being herself a specialty is called in to do evalues. I am in formal just being helpless to help her. In time past I have always been in a play to see it though and help. Now it seems as though I have lost my touch and I can't reach her. She's turning on all the people and I involved. I am to blame in a sense for bringing certain individuals into her life, though not intently. We talked and prayed and then the tormenting spirits are back. They make her seem like and even look like another person. My brother sees it and stays closed up in his room. Now I have prayed again to God for answers and at the end of the prayers the phone rings. It's a cousin that's never called me in her life. We talk about five minutes and then we hangs up, at this point I remember this and calls back and offer her to come out and help me take care of my daughter. They've met before and hit it off pretty good. She immediately agrees and our lives and I drives to Mississippi to pick her up relaxes a bit.

Ira cooks and cleans and keeps her in perfect condition. Now I can work and not worry about her falling again or having to manage the house and cooking. At first she loves it then she feels she's lost her independences completely and resents it. Half of the time she and Ira's at each other and Ira are threatening to leave. It's becoming too, much I try to explain to Ira's she's sick and what's happening with her but without success. So we go on and the back and forth for sometime. By this time Holly's body has weaken to the point she has pressure marks on her and Ira's treats them. She's doing the 3 day week dialyzing still and getting

It's been a long, joyous and hard journey. Today Ms Holly and I are relaxing and just enjoying the time together with each other. Lying there on the sofa facing me with a smile forced though pain and tiredness, she said to me "mom will you help me tell my story"; I said reassuringly, "yes I will", not knowing that our time was limited and I would be keeping my promise 4 years and _____ months late. I am proud to have been chosen to be her mother and have known her for the short time and just to say that through everything she endured she never lost faith in God and who He is. Before leaving the morning of May 3, 2013, she told me that she was tried and that there was something wrong with her body. I knew what was wrong but could not tell her that the doctor had cut her artery and she was bleeding to death. I just wanted her to heal and be alright. She stands up and looks up at me and tells me she needs help, I ask what kind of help and motion for Ira' to get help after pushing the button. I can hear the code blue page as I place her back on the bed. She looks at me and her eyes rolls up and close and her face falls forward. I cup it in my hands and calls to her and her head falls backward and she sigh for the last time. I know it's finished unless God works a miracle. At this moment I feel myself being pushed aside and the team works effectlessly to get her back.

While they are working I leave the room to pray one last time. I wanted God to heal her on this side. But she chose home. Rest Baby Rest is all Mommy can say.

After fought but not with the intend to stay long, just long enough for me to adjust enough to let her go.

www.ingramcontent.com/pod-product-compliance
Lightning Source LLC
Chambersburg PA
CBHW081307180526
45170CB00007B/2599